# Human evolution

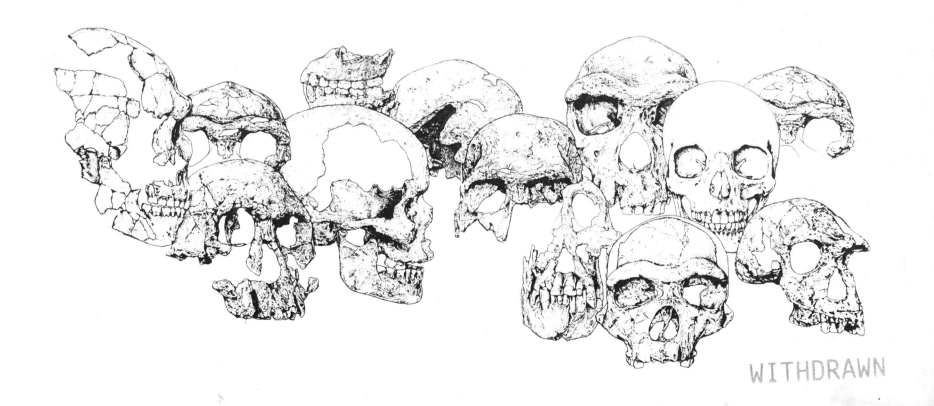

# Human evolution

AN ILLUSTRATED GUIDE BY

*Peter Andrews and Chris Stringer*

PAINTINGS BY

*Maurice Wilson*

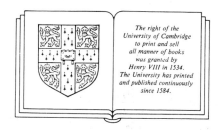

The right of the
University of Cambridge
to print and sell
all manner of books
was granted by
Henry VIII in 1534.
The University has printed
and published continuously
since 1584.

## Cambridge University Press

Published by the British Museum (Natural History),
Cromwell Road, London SW7 5BD and
the Press Syndicate of the University of Cambridge,
The Pitt Building, Trumpington Street, Cambridge CB2 1RP
32 East 57th Street, New York, NY 10022, USA
10 Stamford Road, Oakleigh, Melbourne 3166, Australia

First published 1989

Printed in Great Britain by W. S. Cowell Ltd., Ipswich, England

Library of Congress Cataloguing in Publication Data

Andrews. Peter. 1940–
    Human evolution: an illustrated guide/by Peter Andrews and
    Chris Stringer: paintings by Maurice Wilson.
       p. cm.
    ISBN 0–521–38824–4
    1. Human evolution.    I. Stringer,    C. B. II. Title.
  GN281.A55 1989
573.2—dc20

ISBN 0–565–01020–4   BM(NH) hardback edition
ISBN 0–521–38824–4   CUP paperback edition USA only

# Contents

# Foreword

The illustrations in this book include some of the last works of Maurice Wilson, who died on 1 November 1987 at the age of 73. He possessed an international reputation as a painter of animals.

Maurice Charles John Wilson was born in London in 1914. For most of his life he lived in various places in Sussex and Kent – for the last thirty years or so at Bidborough, near Tunbridge Wells. His artistic education was acquired between 1928 and 1932 at the Hastings School of Art, and then for another three years in London at the Royal Academy Schools. His subsequent career, though mainly as an illustrator, included also some teaching of his artistic skills to others; in the middle-to-late 1940s he taught anatomy and other subjects at the Bromley College of Art and also plant drawing at the Royal School of Needlework.

Maurice was an extremely productive artist, concerned mainly with wildlife; most of his works depict some sort of animal, living or extinct. His animal paintings were done from life wherever possible, often in zoos both in Britain and on the Continent. He was a first-rate draughtsman who observed the animals he drew very carefully and possessed a thorough knowledge of them; he was absolutely convinced that animal reconstructions should be firmly based on anatomical principles. His chosen medium was generally watercolour, but he also used oils, tempera and acrylic; the last, so he said, was best able to withstand the irresponsible practice of certain editors of standing their coffee cups on his finished paintings! As to his style, that was unique and unmistakable; a Maurice Wilson painting needs no signature. I have always felt that much of his work has a slightly Japanese flavour in its line-and-wash technique, its uncluttered backgrounds and its generally ethereal quality, but it was entirely original and in no way derived. Mammals were undoubtedly the creatures that he painted best, and of his mammals I liked his primates – monkeys, apes and Man – best of all.

As soon as Maurice left art school he started to undertake whatever illustrative jobs he could get. He quickly became known in the publishing world, and over the next half-century he illustrated innumerable books, book jackets, magazines and picture-cards – almost always with wildlife subjects. He established a strong connection with the British Museum (Natural History); his best-known works being the reconstructions of fossil Man for Professor Le Gros Clark's anthropological handbooks. The reconstruction of fossils became his particular speciality and were enormously popular throughout the world. Demand for them continues unabated to the present day. Maurice's last major job was the background paintings for the exhibits in the small museum at the Jeffery Harris Memorial Reserve at Sevenoaks.

Maurice was deeply involved with the founding of the Society of Wildlife Artists, of which he later became Vice-President, and he was elected to the Royal Institute of Painters in Watercolour. He exhibited at both those organizations.

Wilson the man was no less characteristic than his paintings. He was kind and generous, friendly and unassuming. I knew him quite well, first through our common membership of the Tetrapods Club (an all-male zoological dining club that met once a month in Central London) and then through our working together in 1970 on a series of Brooke Bond Tea picture cards, 'Prehistoric Animals', for which Maurice did the illustrations and I the text. He certainly had some curious habits. For example, he invariably wore around his neck a red tartan scarf which he absolutely refused to remove, even at semi-formal dinners; it was rumoured that it was permanently glued to his throat. He was often seen in a tatty old raincoat, tied around his middle with a piece of string, and his ancient Ford Anglia (in which he sometimes gave me a lift home from Tetrapods dinners) had to be seen to be believed; yet in both cases he could have afforded better. At first I used to speculate as to whether such endearing Wilsonian idiosyncrasies might be no more than deliberate affectations put on to amuse his friends, but I soon realized that Maurice was indeed a genuine eccentric, incapable of any sort of affectation.

We can no longer delight in Maurice Wilson's company, but we can continue to enjoy the fruits of his artistic endeavours. The paintings in this book – some done years ago, others commissioned specially not long before he died – are among his best. Happy viewing!

Alan Charig
Formerly of the Department of Palaeontology
British Museum (Natural History)

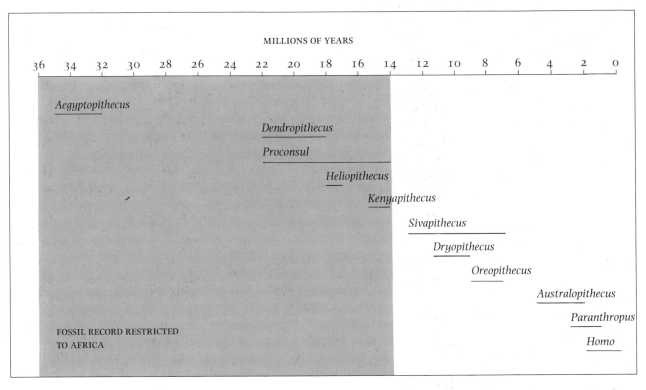

FIG. 1. Approximate dates of the fossils described in the text. From 35 to 14 million years ago hominoids were restricted to Africa.

# Introduction

The story of human evolution stretches back some 30 million years, to a time when luxuriant rain forest covered what is now dry desert in North Africa, and where there lived a rather unremarkable monkey-like animal, called *Aegyptopithecus*. This creature was neither a monkey nor an ape, but it was a primate, a member of the group of mammals to which all present day monkeys and apes, including humans, belong. Known as a pre-hominoid primate, it is considered to be broadly ancestral to all the living monkeys and apes of Africa and Asia. From these earliest beginnings and through various intermediate stages, the story of human evolution is told here by means of paintings and explanatory text. The paintings show the different primates in their natural settings, reconstructed from evidence of fossil plants and animals; the texts give a brief description of each primate, its main adaptations to its environment, how it lived, and particular points of interest especially those relating to differences between groups.

Fossils have provided us with vital clues in the search for the so called 'missing links' which connect the most ancient primates with humans, but before the fossil evidence can be interpreted it is important to discern how the descendants are related. Man's closest relatives in the animal kingdom are other higher primates: New World monkeys (from South and Central America), Old World monkeys (from Africa and Asia), and Hominoidea, a superfamily which includes humans and ancestral prehumans (Hominidae), gorillas, chimpanzees, orang-utans (the Great

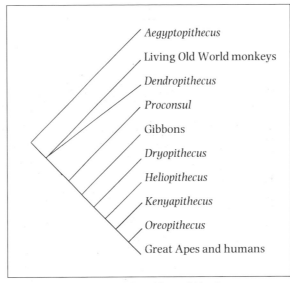

FIG 2a. Relationships of the Old World higher primates.

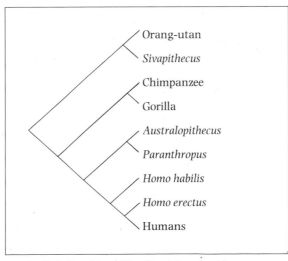

FIG 2b. Relationships of the Great Apes and humans.

Apes) and many species of fossil apes. Our closest living relatives are the chimpanzees and gorilla, with the orang-utan more distantly related. The timing of evolutionary events and the relationships of the fossil primates are summarized here in figures 1 and 2. The period covered spans the time for which fossil evidence for pre-hominoid, hominoid and hominid evolution is available in Africa, Europe and Asia.

After *Aegyptopithecus* (discussed above), the next piece of evidence for the story of human evolution is not until 8 million years later with the fossil hominoid *Proconsul*. This is the earliest known hominoid, identified as such by the adaptations present in its limb bones. The hominoids generally have highly specialized limb bone morphology, whereas their skulls and teeth are rather generalized. Some of these limb bone specializations are seen in the fossilized bones of *Proconsul*. They distinguish it from the monkeys, which retain a more generalized limb bone morphology similar to that of the more primitive hominoid-like primate named *Dendropithecus*. Evidence of both groups has been found in early Miocene deposits of East Africa from about 22 million years ago.

Slightly later in the Miocene, the first indication of the geographic spread of the hominoids beyond the limits of tropical Africa emerges. The existence of *Heliopithecus* from Saudi Arabia, then within the geographical limits of North Africa, attests that hominoids had moved north to the limits of the African continent 17 million years ago, but it was not until more recently, about 13 million years ago, that the first hominoids are recorded

A comparison of the jaws of fossil hominids with a recent chimpanzee and human, showing the contrasts between the jaws in their general shape, thickness and relative sizes of the different teeth. Note in particular the smaller canine and distinctive premolar shape of the hominids, and the large molars in some early hominids, especially *Paranthropus*.

EARLY HOMINOID

FOSSIL HOMINIDS

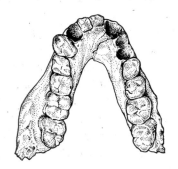

*Sivapithecus alpani* from Turkey

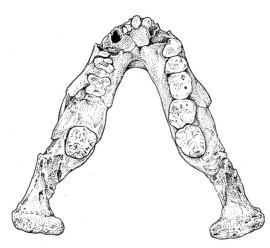

Lucy (*Australopithecus afarensis* – AL-288-1) from Hadar in Ethiopia

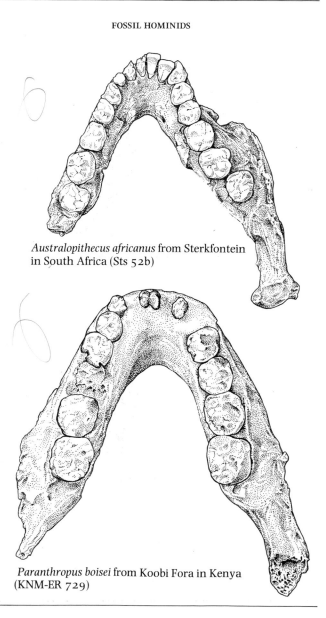

*Australopithecus africanus* from Sterkfontein in South Africa (Sts 52b)

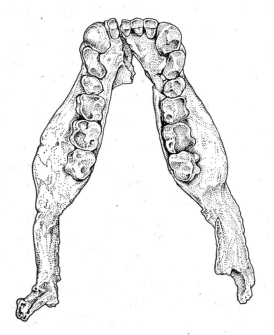

*Sivapithecus indicus* from Pakistan

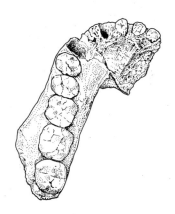

Large australopithecine from Hadar (AL-333w60)

*Paranthropus boisei* from Koobi Fora in Kenya (KNM-ER 729)

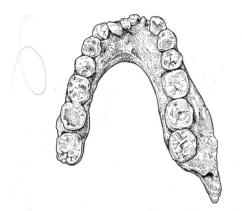

*Homo habilis* from Olduvai in Tanzania (OH 1 3)

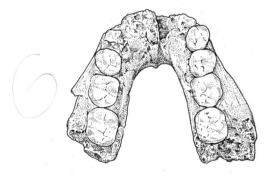

*Homo habilis* from Koobi Fora in Kenya (KNM-ER 1 8 0 2)

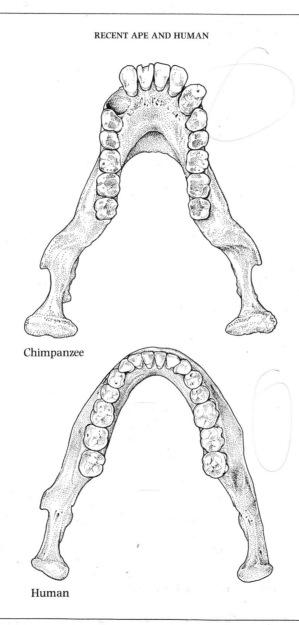

Chimpanzee

Human

outside Africa in the form of *Dryopithecus* in Europe and *Sivapithecus* in eastern Europe and Asia. The latter species appears to be related exclusively to the orang-utan amongst living apes, and may have been ancestral to it, but the other fossil hominoids cannot be related directly to any of the living hominoids. Thus there is no direct evidence for the ancestry of chimpanzees or gorillas, and neither is there any evidence at this stage (10–13 million years ago) for the origin of the line leading to humans (the Hominidae).

Finally, completing the survey of pre-hominid evidence, a rather peculiar hominoid called *Oreopithecus* is discussed. *Oreopithecus* lived 8 million years ago on an island made up of what is now central Italy. Since the fossil material found is relatively complete, more is known about this hominoid than about most fossils, and it seems to be a member of the group comprising of the Great Apes and humans. It thus provides us with useful information about the stage from which the human lineage stepped off.

It was sometime between 5 and 10 million years ago that the ancestors of the gorilla, chimpanzees and humans split from a common stock. There is no direct evidence in the fossil record for the timing of this split, mainly because so few sites are known for this time period in Africa. For example, there is no fossil evidence at all for the origin and evolution of the lineages leading to chimpanzees and gorillas, and it is still uncertain whether one or both of these African apes are more closely related to modern humans. The first good evidence of the evolutionary split comes from fossils of the earliest definite hominids; small brained but bipedal creatures (that is, they walked upright on two legs). The hominids are part of the hominoid radiation, and they are traditionally distinguished from other hominoids (i.e. the apes) at the family level as the Hominidae. This is an anthropocentric practice which in fact gives too great an emphasis on the human lineage, but the term hominid will be used here in this traditional sense. The earliest hominids lived in southern and eastern Africa between 2 and 5 million years ago. These are the australopithecines, a group that was intermediate in many respects between the apes (especially the African apes, chimpanzees and gorillas) and later humans. In other respects, however, they were quite specialized, for example in the shape of the face. They also had greatly enlarged teeth, and although they were clearly bipedal their type of bipedalism probably differed in important respects from our own. An even more specialized subgroup of more robust australopithecines may have persisted until as recently as 1 million years ago.

The earliest and most primitive australopithecines are known from East Africa, and are usually classified within the species *Australopithecus afarensis*, after the Afar region of Ethiopia where the famous skeleton of 'Lucy' was found. Experts are not in fact agreed on whether only one species of australopithecine existed before 3 million years ago, but between 1 and 3 million years ago there were certainly several different types which may have descended from *A. afarensis* or its contemporaries. Between 2 and 3 million years ago the gracile *Australopithecus africanus* lived in South Africa, where it was succeeded by a more specialized robust form called *Paranthropus robustus*, which is not considered to be a human ancestor. In East Africa an even more robust form existed called *Paranthropus boisei*, and a recent discovery from northern Kenya indicates that a robust group existed there 2.5 million years ago.

By 2 million years ago, some early human ancestors had evolved distinctive characters such as a larger brain and a pelvic and limb structure more like that of present day humans, and these changes are recognized as marking the appearance of the genus *Homo*. The first known human species, *Homo habilis*, existed in East Africa and perhaps South Africa from about 2 to 1.5 million years ago, by which time stone tools were certainly in use. By 1.5 million years ago hominids more similar in many ways to living people had evolved; these have been regarded as an early form of the species *Homo erectus*. This species was the first hominid known to have spread beyond Africa, and it is represented by types such as 'Java man' in S.E. Asia and 'Peking man' in China. By 250 000 years ago, humans with even larger brains had evolved, and although opinions differ these are usually regarded as an archaic form of our own species *Homo sapiens*. Fossils from sites such as Broken Hill ('Rhodesian man') and Swanscombe are representatives of this group of 'archaic *sapiens*'. The most recent members of this group are the Neanderthals of Europe and Asia, who existed as recently as 35 000 years ago. Alongside later archaic *sapiens*-like hominids was another group identical in most physical respects to modern humans. This group, first known about 100 000 years ago from Africa and the Middle East probably gave rise to modern humans in Africa and elsewhere, with the archaic hominids becoming extinct. In Europe, for instance, the Neanderthals were replaced by a global expansion of modern *Homo sapiens* (exemplified by the European Cro-Magnons) by 30 000 years ago. Since the arrival of modern *Homo sapiens*, the main evolutionary changes have been those concerned with local differentiation and adaptation, including the evolution of 'races', accompanied by increasing cultural diversity and complexity.

Despite some gaps in the fossil record of human evolution, it is better documented than for many other species of mammal, and the last three million years are particularly well known. We can now see that in the geologically short time span of some 35 million years the hominoids underwent considerable evolutionary radiation into a variety of habitats on the three continents of Africa, Europe and Asia, and through the most versatile of all hominoids, *Homo sapiens*, they are now represented in every habitable part of the world. The current small numbers of other hominoid species give little indication of the past evolutionary success of the group, and it is hoped that this book will help bring to life some of our extinct relatives and ancestors.

# Egyptian apes: *Aegyptopithecus*

More than thirty million years ago, during the period known as the Oligocene, the world was a very different place to the one we know today. Over what are now the deserts of North Africa grew extensive rainforests, and in a place called the Fayum depression in Egypt some of the evidence for the alterations in climate and vegetation can be seen. Piles of massive fossilized tree trunks and the remains of many forest-dwelling animals dating back to the Oligocene period can be found there.

As might be expected, the animals living at that time were also different, and the picture on the opposite page shows one of them that is of particular interest to us because it may be a very early human ancestor. It was a small animal that lived in the trees, rarely coming to the ground, and it is called *Aegyptopithecus*, Latin for 'the ape from Egypt'. It would have looked very much like some of the living monkeys do today, particularly the New World monkeys such as the howler monkeys living in South and Central America, and like them it was a slow deliberate climber. It moved about in the trees on all four legs, feeding on the fruit that could be found all year round in the tropical forests. The males in particular had long dagger-like canine teeth, and these were probably used for breaking into tough kinds of fruit, although they may also have been useful for defence against enemies and for display against other males.

*Aegyptopithecus* is particularly interesting as it is the earliest known representative of the group of higher primates to which humans belong. It is

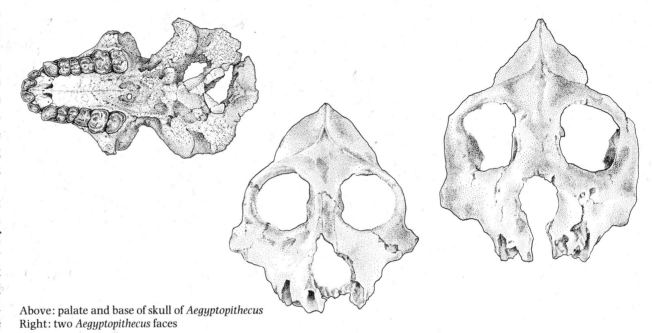

Above: palate and base of skull of *Aegyptopithecus*
Right: two *Aegyptopithecus* faces

clearly related to modern monkeys and apes, because it shares important characters with them, such as the growth of bone around the eyes and the reduction in numbers of teeth. It lacks any characters which are diagnostic of these two groups and so appears to be related to both equally. In other words, it lacks hominoid specializations, and so is not an ape, and it lacks cercopithecoid specializations and so it is not a monkey. In fact, it probably lived at a time before the monkeys and apes of the Old World had evolved. It would be a mistake, however, to think of *Aegyptopithecus* as a very primitive primate. In its way it was just as specialized and successful as the living primates are today. It had to live in competition with other animals, including at least four other primates present in the same habitat and a great variety of rodents, carnivores and hyraxes, which may have competed with it for food or preyed upon it. It is likely to have lived in groups with a complex social structure, since this is a characteristic of all living higher primates, and in fact if it were still living today it would not stand out as unusual in any respect. It was, therefore, the first of the higher primates with modern characteristics, and it must stand close to the ancestry of all later higher primates, including humans.

*Aegyptopithecus zeuxis* – Egyptian apes. The forerunners of both monkeys and apes, these small arboreal primates lived 35–32 million years ago in tropical rainforests of northern Egypt. ▷

# African wood apes: *Dendropithecus*

Some 10 million years after *Aegyptopithecus* lived another very similar primate, pictured here. This is called *Dendropithecus*, the wood ape or the ape that lived in trees, and it lived in East Africa 20 million years ago, during the Early Miocene, in the area now known as Kenya. Here also the country was very different from the way it is now. East Africa today is dominated by the rift valleys and their associated volcanoes, not all of which are extinct. On the high lands where there is plenty of rain, thick forests grow, while the low lands in between, where the rain is intercepted by the mountains, are covered by open grasslands and semi-arid deserts; the famous savannas of eastern Africa. Back in the Miocene, however, the mountains and rift valleys had not yet formed, so that the land was flat and the climate and rains distributed more evenly. In this situation, much of the land was covered by tropical forests in which a great variety of primates and other animals lived.

*Dendropithecus* was a similar type of primate to *Aegyptopithecus*, even though it occurred much later. Its similarity is partly due to the fact that the species are closely related, and partly because both species lived in such similar environments. *Dendropithecus* was also a small, fruit eating primate, living in trees in which it was probably more agile than *Aegyptopithecus*. Because their forest environments were similar, the forces of natural selection acting on the two primates would have been similar, and as a result the two species closely resemble each other.

There are a few ways in which *Dendropithecus* differs from *Aegyptopithecus*, and these provide interesting information on primate evolution. Whereas *Aegyptopithecus* had a projecting dog-like face, similar in many respects to the faces of the very early primates, the prosimians, *Dendropithecus* had a shortened face, more like the living monkeys and apes. In addition, the arm bones of this latter primate, particularly the elbow joint, are more like those of monkeys and apes, and these changes show that *Dendropithecus* is more closely related to the monkeys and apes than *Aegyptopithecus*. In fact, it is clear from this evidence that *Dendropithecus* can be classified with the monkeys and apes and must be close to their ancestral condition.

We have seen how *Aegyptopithecus* of the Oligocene period was probably a very early ancestor of the Old World monkeys and apes, and this tells us that this group of higher primates originated at least 30 million years ago. *Dendropithecus*, which lived 22 million years ago shows us that by that time we are close to the actual divergence between monkeys and apes, because so many of the shared characteristics of the two groups had developed by then.

The early evidence of both monkeys and apes is restricted to Africa, and it seems likely on this basis that they evolved on this continent. Fossil monkeys are not known until later, about 18 million years ago, and it is probable that they were living in different habitats from the hominoids and so have not been preserved in the same fossil deposits. Monkeys became specialized in their teeth and jaws, but at this early stage of divergence from the hominoids it is probable that they were still rather similar in morphology and difficult to distinguish on the basis of fragmentary fossils.

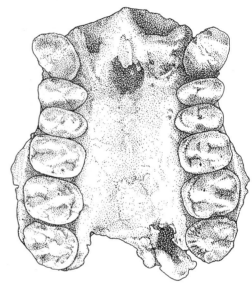

Upper jaw of *Dendropithecus*

*Dendropithecus macinnesi* – African wood apes. Lived in tropical rainforests of East Africa about 20 million years ago. ▷

# Earliest apes: *Proconsul*

Living at the same time and in the same places as *Dendropithecus* was a group of apes called *Proconsul*. This is a diverse group that was widespread in eastern Africa, where it survived for many millions of years. It was named in 1933 after a famous chimpanzee called Consul which was then living in London Zoo, and at that time it was thought that this group of fossil apes was ancestral to the chimpanzees, hence Pro- (before) Consul. It includes at least five species living in eastern Africa between 22 and 14 million years ago, and these species ranged in size from that of small monkeys up to gorillas.

Some of the *Proconsul* species lived in tropical forests, but it is quite probable that others did not. The largest species is found only in sites associated with tropical forest animals whereas some of the smaller species are found in more mixed environments. For the first time, therefore, there is some evidence of primates living outside the tropical forests that seem to be their natural home. This is particularly interesting when it is considered that *Proconsul* is the earliest recognized hominoid ape, that is a primate belonging to the group to which humans and the living apes belong. We have seen how *Aegyptopithecus* and *Dendropithecus* are related, and may be ancestral, to the living apes and humans, but *Proconsul* is an example of an actual member of this group.

Superficially, *Proconsul* would have looked much like chimpanzees do today, but it differed greatly in detailed morphology. For instance, chimpanzees have long arms, short legs, projecting faces with large canines, and a curious method of walking by supporting their weight on the knuckles of the hands. None of these adaptations were present in *Proconsul*, which had a lighter build, legs and arms of equal length, strongly gripping hands and feet and a shorter, rounder face. It moved on all fours through the trees, and probably often hung below the branches which it gripped with both hands and feet.

The identification of *Proconsul* as an early hominoid shows that this group originated at least 22 million years ago. This means that the group to which humans belong has been in existence for at least that long, and maybe several million years longer. It is also interesting to note that whereas the living apes are rather rare and widely scattered in Africa and Asia, these fossil apes were more common and apparently successful. If all the early Miocene apes or ape-like primates are considered together, there were often five or six species present at one place and one time, whereas today there are never more than two species of living ape found together. Instead there are many species of monkey, which seem to have replaced the apes in most of the tropical forests of the Old World. This shows that the apes, having been a successful and vigorous group 15 to 20 million years ago, have declined in importance and today are only represented by a few relic populations. Man is one species that has reversed this trend.

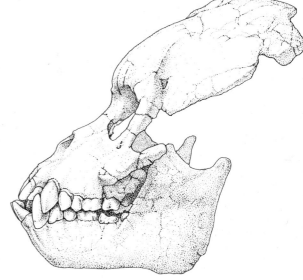

Skull of *Proconsul*

*Proconsul africanus* – The earliest known true apes; like *Dendropithecus*, they mostly lived in tropical forests in East Africa. This is the smallest *Proconsul* species, which was associated with more open country. ▷

# Sun apes: *Heliopithecus*

At a later stage in the Miocene of Africa a significant development occurred in hominoid evolution. This was the development of a whole new dietary modification resulting from the thickening of the enamel on the surface of the teeth, completely changing their potential function. Enamel forms a protective hard coating which prolongs the life of the teeth. For animals in the wild, which depend on their teeth to process all their food, their very survival depends on having a good set of teeth, and when their teeth wear out they are unable to continue eating, and die. By increasing the thickness of the enamel the life of the teeth and hence the potential life of the animal may also be increased.

The first evidence for this increase in enamel thickness is seen in a group of hominoids that is represented here by *Heliopithecus*. The name comes from the Greek word for the sun, 'helios', and the group was so named because the original specimen was found in the Arabian desert. *Heliopithecus* occurs in deposits dating back 16 to 17 million years, and its teeth show the early stages of enamel thickening. This trend culminates in later hominoids of the genus *Kenyapithecus*, which lived in East Africa 14 to 15 million years ago, the specimens of which have a greatly thickened layer of enamel on their teeth.

*Heliopithecus* and *Kenyapithecus* have other similarities which suggest that they were members of the same lineage. Although the specimens found so far are very fragmentary, it is apparent that details of the molars and premolar proportions are common to these two genera and are not found in any other hominoid primate. Thus the evidence points to *Kenyapithecus* being a more advanced derivative of *Heliopithecus*, both in terms of morphology and of function.

*Heliopithecus* and *Kenyapithecus* lived in habitats that were less benign than the tropical forests of earlier hominoids. These later hominoids lived in tropical and subtropical woodlands, present even in Saudi Arabia in areas which are now desert, and this is shown in the accompanying picture. As today, woodlands such as these were more seasonal than the tropical forests of the earlier hominoids, and there would have been longer periods during which food was relatively hard to come by. Like the winter in northern latitudes, the dry season of the tropical woodlands is a time when life shuts down, food resources are stored, and animals foraging on a day to day basis have difficulty in finding enough to eat. The added advantage of an adaptation like thick enamel could have been a great asset in these hard times, increasing the variety of foods available to the hominoids.

Brief mention can also be made here of a new fossil ape just described from East Africa. This is named *Afropithecus* and it is both similar in morphology to *Heliopithecus* and *Kenyapithecus* and probably lived in similar habitats. These three genera show the great range and diversity of the hominoids during this period of the Miocene.

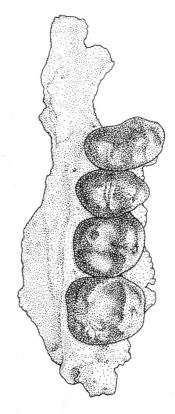

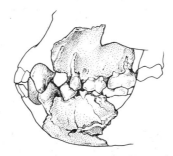

Part of jaw of *Kenyapithecus* (reconstructed)

Part of jaw of *Heliopithecus*

*Heliopithecus leakeyi* – sun apes from Saudi Arabia. Lived 17 million years ago in woodland which is now part of the Arabian Desert. These apes show the first evidence of a new hominoid adaptation; thickened enamel on their teeth. ▷

# European wood apes: *Dryopithecus*

Until about 13 million years ago, hominoids apparently occurred only in Africa. We have seen how *Proconsul*, *Dendropithecus*, *Kenyapithecus* and *Heliopithecus* all lived there between 14 and 22 million years ago, and although fossil sites of similar ages in Europe and Asia have been extensively studied, no sign of these or any other hominoid primates has been found. This suggests that the hominoid group originated more than 22 million years ago in Africa and it did not reach Europe and Asia until some time later. The time of this dispersal is indicated by two fossil groups which are the subjects of this and the next section.

*Dryopithecus* was a fossil ape that lived in western and southern Europe during the period 9 to 11 million years ago. The first specimen was found in the south of France near the village of Saint Gaudens in 1856, and because it was found associated with oak leaves it was named after the dryads, the oak nymphs of Greek mythology. Later, further finds were made in France, and more recently collections of the same or closely related species have been made in Hungary and Spain.

Many aspects of its teeth and skull show that *Dryopithecus* was still quite similar to the earlier hominoids like *Proconsul*. It differed in several important respects, however, particularly in its limb bones. The bones of the elbow joint are more advanced than in *Proconsul*, and are very like those of the Great Apes and humans. The skull also has characters that are more advanced than the ancestral hominoid condition, like robust ridges over the eyes. Analysis of these characters

allow us to suggest that *Dryopithecus* is more closely related to the Great Apes and humans, than for instance, the living gibbons are. However, in several important characters it is less

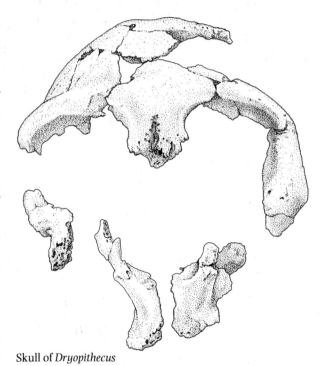

Skull of *Dryopithecus*

advanced than *Kenyapithecus* and *Heliopithecus*, discussed in the last section, so that even though these lived earlier in time they are thought to be closer to the ancestry of living apes and humans than is *Dryopithecus*. It seems likely that the living apes and humans evolved in Africa, and *Dryopith-*

*ecus* should be viewed as an emigrant hominoid from Africa, which did not evolve into any species living today.

The adaptations *Dryopithecus* had to its environment are particularly interesting. Its teeth and jaws are similar to those of *Proconsul*, and it may be inferred from this that it had a similar diet of fruit. The extra buttressing of the face implied by the larger brow ridges, and the presence of a different kind of buttressing of the lower jaw show increased resistance to strain imposed by chewing, so that it is likely that its diet was coarser, or harder, than that of *Proconsul*. This is not surprising, since it lived in the south temperate to subtropical woodlands of Europe where the climate is more seasonal than *Proconsul*'s home in the tropical African forests, bringing a greater reliance on lower grade food items during winter. The European forests would have had many familiar European tree species like oaks and pines in addition to more exotic subtropical species. The group was probably still mainly arboreal, as shown here, but *Dryopithecus* was a more heavily built animal than any of the *Proconsul* species and probably moved about on the ground as well.

*Dryopithecus fontani* – European wood apes. Widely distributed in western and eastern Europe 9 to 11 million years ago. This reconstruction is based on relatively complete material from Hungary.  ▷

# Swamp apes: *Oreopithecus*

Paradoxically, *Oreopithecus* is both one of the best known and one of the least understood fossil primates. Originally described in 1872, it was one of the earliest discovered fossils, and the numerous skulls, jaws and skeletons discovered since then mean it is probably the most completely known of all fossil apes in terms of material available for study. Despite this, *Oreopithecus* is an enigma over which scientists continue to argue, partly due to its strange morphology and partly to its potential importance to the general understanding of hominoid evolution.

*Oreopithecus* is known only from late Miocene lignite deposits in Italy. These are dated at between 7 and 9 million years, and they were formed in an extensive area of swamp forest in what may have been a large island, comprising much of present day central Italy. The swamp was an area of stagnant waters, many small islands and marshy areas, and it was heavily vegetated with extensive subtropical forests that were preserved in the swamp waters as lignite. A great thickness of lignite had built up over the years, and when it was being mined to be burnt as a low grade fuel, *Oreopithecus* and other animal remains were found, most of them compressed and distorted by the weight of the sediments on top. One of these skeletons is shown in the accompanying diagram.

In such an environment, living in trees and keeping off the marshy ground was obviously a good adaptation, and *Oreopithecus* was a specialized tree dweller. It had long arms and hook-like hands which enabled it to hang from branches of trees in a way rather like the sloths of today, and it might also have resembled sloths in the way it moved and fed in the trees. Its adaptations for this arboreal life are similar to those of the living Great Apes, which also have the ability to hang from branches, although this is probably not their primary adaptation. In contrast, *Oreopithecus*'s more monkey-like teeth were highly specialized for feeding on leaves, and it is the contrast of the monkey-like teeth and ape-like postcrania that has given rise to the controversy over its relationship with other primates.

It is possible to trace the ancestry of *Oreopithecus* back to the early Miocene, to hominoids contemporary with *Proconsul* and *Heliopithecus*. *Oreopithecus* itself, however, strongly parallels the Great Apes and humans in its morphology, and it probably branched off from the hominoid lineage before the orang-utan and the African apes and humans diverged. It thus provides an additional link in this little known part of the line leading to humans.

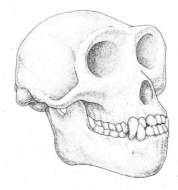

Reconstructed skull of *Oreopithecus*

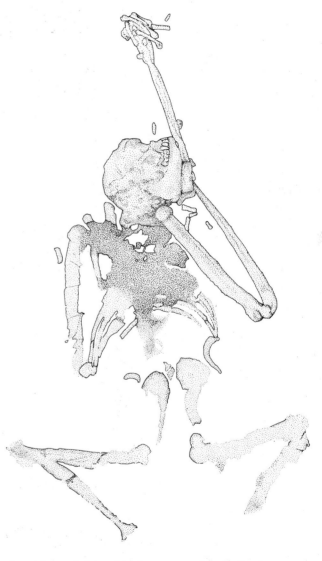

*Oreopithecus* skeleton found compressed under lignite deposits

*Oreopithecus bambolii* – swamp apes. Very complete fossil remains have been collected in Italy in deposits 7–9 million years old. It is peculiar in its rather monkey-like teeth and its specializations for living in swamp forests. ▷

# Thick enamelled apes: *Sivapithecus*

In the section on *Heliopithecus* and *Kenyapithecus*, the changes brought about by thickening the enamel of the teeth were discussed. Like these groups, *Sivapithecus* also has thickened enamel, and its dietary adaptations would have been very similar. In most other respects, however, it is very different.

*Sivapithecus* is best known from later Miocene deposits of Greece, Turkey and Indo-Pakistan, dating back from 7 to 11 million years. Included with it here is another group of fossils called *Ramapithecus*, once thought to be distinct, but now recognized as belonging to the same group. The name *Sivapithecus* has priority over *Ramapithecus* because it was established in the scientific literature first, so this is the name that is used. Their names were derived from the Indian gods Rama and Siva, as the first fossils of this group were found in India, but the most significant recent finds have come from Pakistan and Turkey. From the latter there is a very large collection dating back as far as 13 million years, which provides both the earliest record of hominoids living outside Africa, and the first record of *Sivapithecus*.

Recent finds of *Sivapithecus* (including *Ramapithecus*) provide strong evidence for shared ancestry with orang-utans. This relationship is indicated by the many characters of the face, nose and palate that are uniquely shared by orang-utans and *Sivapithecus*, while being absent from other primates. It was previously thought that *Ramapithecus* was an early hominid, ancestral to humans, but since the orang-utan shares ancestors

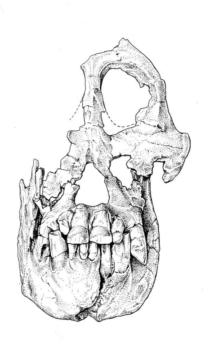

Face of *Sivapithecus* from the Potwar Plateau in Pakistan

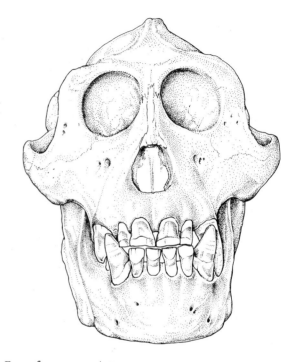

Face of an orang-utan

with this group and is only distantly related to humans, this is no longer a reasonable suggestion, although many textbooks on human evolution have yet to recognize this fact.

*Sivapithecus* was a relatively large hominoid, and apparently there were considerable differences in size between the sexes, as there are in the orang-utans today. It lived at similar latitudes to *Dryopithecus*, but never at the same place and time, suggesting that the two groups had different requirements that kept them separate. *Dryopithecus* lived in warm temperate to subtropical woodlands of southern Europe in the Miocene, and *Sivapithecus* occupied more open and probably more seasonal woodlands. In this sort of open woodland it would not be possible for an animal the size of *Sivapithecus* to live entirely up in the trees, and it seems likely that it lived part of the time on the ground, as shown here. It still had limbs mainly adapted to arboreal life, however, and probably never moved far from the protection of the trees.

*Sivapithecus indicus* (including *Ramapithecus*) – thick enamelled apes from Turkey and Indo-Pakistan. ▷
Thickening of the enamel on the molar teeth during hominoid evolution is carried to its extreme
in this group, which is also distinguished by its similarities to the living orang-utans.

# The southern ape of Afar: *Australopithecus afarensis*

Sometime before 4 million years ago, the evolving lines leading to the African apes split from the line leading to humans, the hominid line. From 4 million years ago the hominid line was clearly established and recognizable in the form of creatures which were bipedal, that is they walked upright on two legs, but which in many other respects still retained hallmarks of that last common ancestor shared with the African apes. This earliest known hominid species is called *Australopithecus afarensis*, the southern ape of Afar; the name results from the location of the finds, the genus being previously known from South Africa and the species from the Afar region of Ethiopia as well as from Tanzania and Kenya.

The remains of *A. afarensis* have so far been found at Laetoli, not far south of Olduvai Gorge, Tanzania; at sites in Ethiopia (most notably at Hadar); and in 1984 a jaw fragment was discovered at Baringo, Kenya, which probably also represents the species. This last find is probably the oldest, since it may date back 5 million years, but other material including the famous skeleton of 'Lucy' from Hadar may be only 3 million years old. With the evidence from Hadar and Laetoli we can build up a picture of *A. afarensis* as a species still very ape-like in many features of the skeleton and probably also in behaviour. Their brains were similar in size to those of apes, although it is disputed whether the shape of the brain did or did not show human characteristics. The projecting face and the jaws were large and ape-like. The jaws housed relatively large back teeth covered in thick enamel, and canines and premolars which,

in some respects, were intermediate in form between those of apes and humans. The skulls of the big individuals, probably males, carried crests like those in gorillas or large chimpanzees, which gave enlarged surface areas for muscle attachments from the jaws and neck. However, despite the ape-like features, *A. afarensis* definitely regularly walked upright, unlike apes, as is established not only from evidence of hip and limb bones from Hadar but also from a remarkable series of footprints preserved in volcanic ash at Laetoli.

The males were probably about 1.5 metres tall, and as heavy as European males of today (up to 70 kg) but there were also much smaller specimens such as 'Lucy' who was less than 1.2 metres tall and could have weighed less than 30 kg. The sheer variation between the big and small individuals of *A. afarensis* has led some scientists to question whether only one species is represented by the fossil material. Other workers believe that the smaller females could have behaved slightly differently to the larger males, accounting for some of the skeletal differences, for example in their greater use of trees for feeding, resting, sleeping or escape. Most of the experts are agreed, however, that although the fossil material and footprints clearly point to creatures which regularly walked upright, their behaviour would have been very unlike that of humans. They were probably ape-like in the extent to which they climbed trees and in their diet which probably still consisted mainly of vegetable materials, especially fruits and leaves, rather than meat. Their use of tools was probably no more developed than that of

modern chimpanzees, and their language and social systems must have been quite unlike those of later true humans.

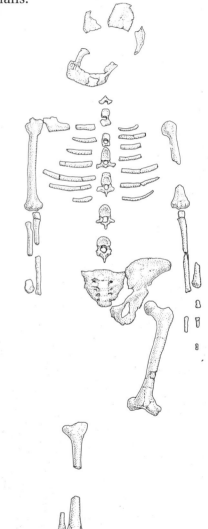

Skeleton of 'Lucy'

*Australopithecus afarensis* – primitive African hominids living 3 to 5 million years ago. Despite the facial resemblances to apes they regularly walked upright, although the retained the ability to cli

# The southern ape of Africa: *Australopithecus africanus*

In the early years of this century, the search for the supposed 'missing link' between ape and man was concentrated in Europe and the Far East, where finds such as 'Piltdown Man' and *'Pithecanthropus erectus'* (Java man) had been made. Thus little notice was taken of a claim by Raymond Dart for a new kind of 'man-ape' from South Africa. The claim was based on a single skull of a young 'ape' which had been found during mining work at Taung, South Africa, in 1924. According to Dart, the shape of the skull and probable shape of the brain, the form of the teeth and jaws, and the inferred man-like posture of the skull all indicated a creature more 'human' than any known ape. Dart worked in the area for many more years before he and another researcher, Robert Broom, started to discover supporting evidence for his earlier claims, in the form of more skulls, jaws, teeth and other bones from additional cave sites. Eventually it became clear that there were actually two types of 'man-ape' from the South African caves. One was a lightly built form called *Australopithecus africanus* (the southern ape of Africa), first named by Dart for the Taung fossil in 1925; the other was a more heavily built form with larger skull, jaws and teeth named *Australopithecus robustus* (robust southern ape) or *Paranthropus robustus* (robust near man), discussed on p. 32.

The species to which the Taung specimen belongs is mainly known from sites dating back between 2.5 and 3 million years. A picture has gradually been built up of the likely appearance of *A. africanus*, although there is much disagreement about the way of life of this creature and its place in human evolution. This bipedal species of early hominid was probably less than 1.5 metres tall and weighed about 30–40 kg. Like *A. afarensis*, the males would have been larger than the females, although the difference in size was not as great as that inferred for the earlier species. The skull of *A. africanus* was similar to that of *A. afarensis* in its combination of ape-sized brain and large projecting face, but there are detailed differences, *A. africanus* having a flatter face and no skull crests. Its teeth may have been adapted for a harder, more abrasive diet with less fruits and leaves, and more gritty roots, tubers and seeds. Despite the fact that there are no skeletons of *A. africanus* as complete as 'Lucy', experts believe the two species were probably quite similar in overall form, although the South African species was less ape-like in certain respects.

Dart also argued for the existence of a primitive 'osteodontokeratic' culture in the australopithecines, a cultural stage when bones, teeth and horns of other mammals were utilized by these early hominids before they had the ability to make adequate stone tools. Dart believed that *Australopithecus* was a carnivore able to hunt other large mammals, and that caves represented the actual homes of the man-apes to which they carried their prey to eat it. Although such an osteodontokeratic cultural stage is certainly theoretically possible, most scientists now believe the evidence for such a culture has not been demonstrated in South Africa. Instead, carnivores such as the leopard and hyena may have been responsible for the

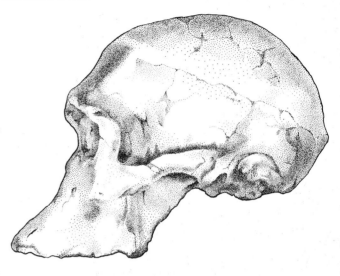

Skull of *Australopithecus africanus* from Sterkfontein in South Africa (Sts 5). Probably about 2.5 million years old

accumulations of bones, teeth and horns, and the australopithecines themselves may have been victims, rather than predators, of other mammals.

The evolutionary position of *A. africanus* is also now uncertain. Some experts believe the species was ancestral to the robust australopithecines, but not to humans. These workers instead prefer to place *A. afarensis* in the position of last common ancestor for both the South African australopithecines and early human groups. Dart's find is now almost universally regarded as a true hominid, as he first claimed in 1925, but its position as a possible human ancestor is likely to be disputed for many years to come.

*Australopithecus africanus.* The extent to which these gracile South African hominids used stone and wooden tools or ate meat is still unknown. The group in the background are driving some hyenas from a kill.  ▷

# Robust near man: *Paranthropus*

As discussed in the last section, the work of Dart and Broom established the existence of very early hominids in South Africa, and it was gradually realized that at least two types of creature were represented: the gracile *A. africanus* discussed earlier, and the larger form which we will call *Paranthropus robustus*. At first it was not clear if the two types could have coexisted as they were discovered at different sites, but eventually it was established that the gracile forms, dated at about 2.5 to 3 million years old, existed before the robust ones, which lived about 1.5 to 2 million years ago. As we have seen, it has also been suggested that the gracile species was ancestral to *Paranthropus*. There may have been an environmental trend towards drier, more open conditions across this time period which may help to explain some of the evolutionary changes, including the development of features of the jaws and teeth of *Paranthropus* which relate to the powerful grinding of hard food. Alternatively, new discoveries of *Paranthropus* from Kenya, dated at 2.5 million years, may provide the ancestor for the later robust forms of both East and South Africa.

Although no partial skeletons of *Paranthropus* have yet been discovered, it was undoubtedly a much larger animal than both *A. africanus* and the small individuals of *A. afarensis* such as 'Lucy'. Female *Paranthropus* probably weighed about 40 kg and males, with a standing height of over 1.5 metres, perhaps twice as much. The skulls of *Paranthropus* look superficially gorilla-like because the large males possessed crests on the top and back of the skull to maximise areas for muscle attachments, particularly for those of the powerful jaws. The brain was the size of a gorilla's, but the peculiar flattened face, disproportionately tiny front teeth and large back teeth, and the rather human-looking base of the skull could never be confused with those of a gorilla.

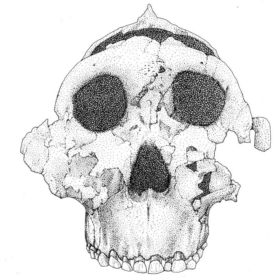

Facial view of the reconstructed skull of '*Zinjanthropus*' (OH5)

These peculiar features of the skull, jaws and teeth reached an even greater extreme in the *Paranthropus* species of East Africa, most famously represented by the skull originally called '*Zinjanthropus boisei*', the East African man. This remarkable specimen was discovered by Mary Leakey at Olduvai Gorge in Tanzania in 1959, and was named after Charles Boise, a benefactor of the Leakeys. Its discovery acted as a catalyst for much subsequent work in the area by the Leakeys and many others since. The impact of the discovery was even greater when, shortly afterwards, new radiometric dating techniques showed that '*Zinjanthropus*' was nearly 2 million years old. This was the first important application of this method to date a hominid fossil, and gave a much greater age than had been expected, but one which is now almost commonplace for early hominids. The fossil itself was the skull of a near-adult individual nicknamed 'dear boy' by the Leakeys and 'nutcracker man' by the media, on account of the enormous back teeth it possessed. Although most experts do not accept '*Zinjanthropus*' as a true human, the last nickname is not entirely inappropriate, given the powerful grinding action of the jaws of *Paranthropus* species.

When '*Zinjanthropus*' was found at Olduvai, the Leakeys at first believed they had found the manufacturer of the primitive 'Oldowan' stone tools which had been found at various sites in the gorge, since the skull lay on a supposed 'living floor' with such tools. Shortly afterwards, however, when the more human *Homo habilis* fossils were discovered in the same levels (see p. 34), the interpretation changed on the assumption that the more advanced hominid was the real maker of the Oldowan industry. A similar story has emerged at Koobi Fora, Kenya, where *Paranthropus* and *Homo* both occur in levels containing stone tools. The possibility still remains, nevertheless, that *Paranthropus* could have used simple tools to help in the discovery and processing of vegetable foods.

*Paranthropus*. This robust vegetarian australopithecine had a long and flat face. The reconstruction is based on the famous '*Zinjanthropus*' skull found at Olduvai in 1959 which is nearly 2 million years old.

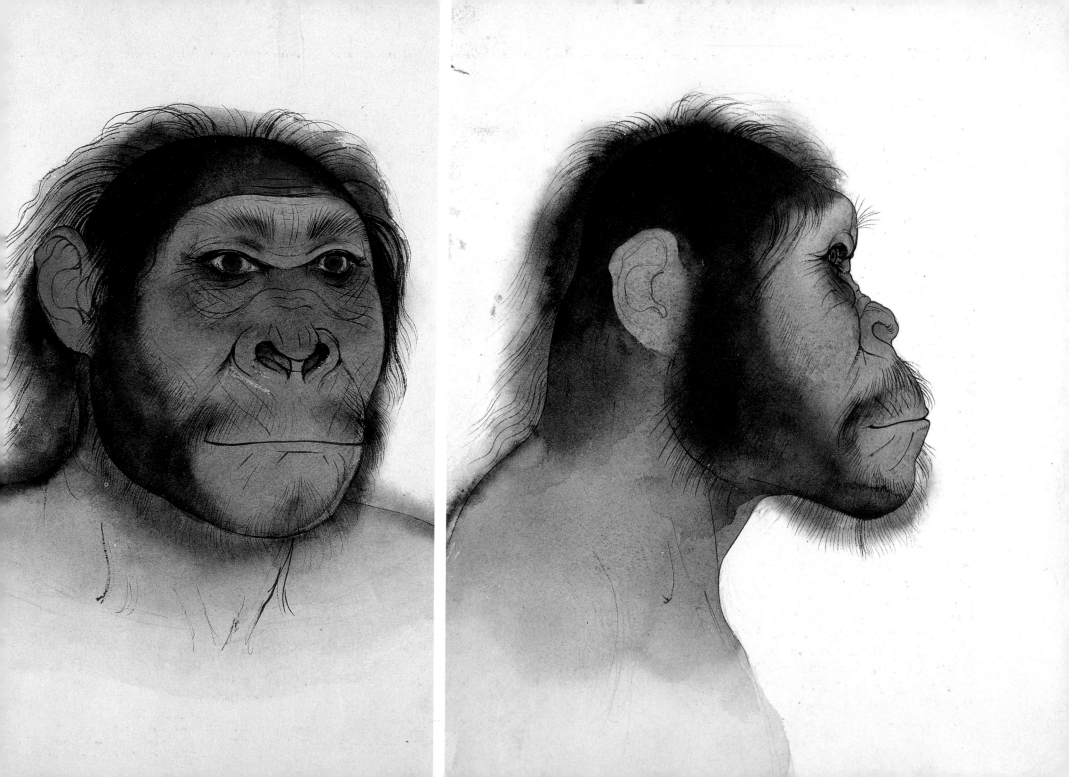

# Handy man: *Homo habilis*

About 2 million years ago the first real humans appeared on the Earth. They are represented by fossils from East and South Africa assigned to the species *Homo habilis*, handy man. Establishing them as 'Man' is not easy, because so many features which we recognize as being uniquely human are behavioural and do not fossilize, e.g. our well-developed social systems, our long period of infant dependence and care, and our methods of acquiring food. Other physical distinctions which today clearly separate us from the apes must have been far less developed 2 million years ago, e.g. our large brain, puny skeleton, small face, jaws and teeth. Nevertheless, experts believe that enough of these special human features can be recognized in *Homo habilis* to warrant this classification as human, although the fossils assigned to this earliest known species of Man vary in the extent to which they show the characters concerned.

The first fossils of *Homo habilis* were found by the Leakeys at Olduvai Gorge between 1959 and 1961. They included jaw bones which appeared to have smaller and narrower teeth than those of the australopithecines; skull fragments indicating an enlarged brain, and some other bones of the hand, leg and foot which are less easy to classify. There was a great deal of argument when the new species was created in 1964 because some scientists thought there was no justification for the naming of a new species when the fossils resembled existing specimens of *Australopithecus* or the early human species *Homo erectus* (discussed on p. 36).

The idea of a distinct human species intermediate in various ways between *A. africanus* and *H. erectus* gained wider acceptance with the discovery of more material not only at Olduvai, but at Koobi Fora, Kenya. The famous discovery in 1972 of '1470 Man', so called because of its

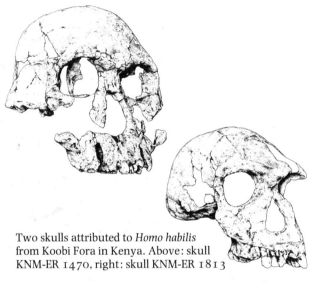

Two skulls attributed to *Homo habilis* from Koobi Fora in Kenya. Above: skull KNM-ER 1470, right: skull KNM-ER 1813

Kenya National Museum number KNM-ER 1470, showed clearly that the brains of some hominids about 2 million years old were larger than those of any australopithecines, yet they were still different from fossils representing *H. erectus*. The brain size of the Koobi Fora skull (750 ml) was about 50% larger than that of a typical australopithecine, and more than half the average size today. Yet 1470 still possessed some australopithecine features in its large, transversely flat but projecting face, and a big upper jaw, which must have

housed large teeth. Other bones of the skeleton found in the same levels as 1470 suggest that these Koobi Fora early humans were as big as many people alive today, although not as large as typical Europeans. In some respects, at least, their hip and leg bones were very similar to our own, but skeletal material from Olduvai is more equivocal, and more like that of australopithecines.

While many scientists accept 1470 as a Kenyan example of *H. habilis*, they are not so sure about other more recent fossils from Koobi Fora. There is another skull (KNM-ER 1813) whose face, upper jaw and teeth look more human than those of 1470, yet the brain size, 510 ml, is no bigger than in gorillas or many australopithecines. A third skull also has more human looking teeth, but quite a small brain (580 ml) and crests on the skull which superficially resemble those of an ape or *Paranthropus*. These other skulls could represent a variant of *H. habilis*, but the variation might be so great that they belong to yet another species of early man still to be named.

The situation is also complicated when it comes to assessing the behavioural evidence from the time of *H. habilis*. Simple tools are present at Olduvai and Koobi Fora, but as they are usually found at sites where there is no evidence of hominid bones it is not certain whether *H. habilis* was the only tool maker living there. Also, despite the evidence of a difference in teeth and jaws from the australopithecines, experts are not sure how dependent *H. habilis* had become on meat-eating and regular hunting for food.

*Homo habilis*. These first true humans lived 2 million years ago. They had a varied diet, and used stone and wooden tools in their search for food. The male in the foreground has the characteristic flat but projecting face of the 1470 skull from Koobi Fora. ▷

# Man at the dawn of the Ice Age

About 1.6 million years ago, soon after the beginning of the Pleistocene, early humans radiated out from their ancestral home to inhabit the tropical and subtropical areas of Asia, and subsequently colonized the more temperate parts of Asia and Europe. As the pace of climatic changes quickened, ice-caps periodically extended from the polar regions and from high mountain ranges, while environments near the equator regularly fluctuated between tropical forests and drier grasslands. At first people must have been driven away or died as a result of such rapid environmental changes, but slowly they learned to adapt to the new climatic conditions and were able to continue living in Europe and Asia during the Ice Ages. The earliest human species to achieve this wide range and adaptability was *Homo erectus*, whose fossil remains were first discovered in Java in 1891, formerly called '*Pithecanthropus erectus*'; the erect ape-man. The remains not only cover a wide geographical range but also span over a million years in time, from the earliest African fossils which might represent this species which are over 1.5 million years old, to the youngest finds from China and Java, which are less than 300 000 years old. Not surprisingly, the finds show a lot of variation, some of which probably represents 'racial variation' due to climatic adaptations or to isolation, and some of which is due to broader evolutionary changes through time such as a general increase in brain size.

The oldest known *H. erectus* remains are probably those from Koobi Fora and West Turkana in northern Kenya, discovered by teams under the direction of Richard Leakey. Koobi Fora has produced *erectus* skulls with brain sizes of about 825 ml, larger than any *habilis* specimen, but few bones of the rest of the skeleton. This deficiency was more than made up for by the discovery in 1984 of a nearly complete skeleton of an *erectus* youth, perhaps 12 years old, on the opposite side of Lake Turkana from Koobi Fora, in deposits similarly dated at about 1.6 million years old. His spinal column looked distinctive and his hip and thigh bones look different from ours, not only in their robusticity and evidence of great muscularity, but also in the way in which they must have operated during body movements. His skull, like those found at Koobi Fora, already displays the distinctively large brow ridge which is a hallmark of human evolution for most of the succeeding Pleistocene up to the time when people like us evolved. Compared with many of the *habilis* fossils the face, jaws and teeth were smaller, but there were also indications that the skull was evolving a longer, lower shape and the bones of the skull and limbs were becoming thicker. This trend is most obvious in the later Far Eastern *erectus* fossils from Java and China.

We believe that all the *H. erectus* people used stone tools. To begin with they probably made tools of the Oldowan type, but some of their descendants developed larger standardized tools which we call handaxes. These remained an essential part of the tool kit for much of the Pleistocene.

While the early *erectus* peoples foraged for meat

Skeleton of an early *Homo erectus* boy from West Turkana, Kenya (WT-15000)

or vegetable foods near the shores of the ancient Turkana and Olduvai lakes they may have encountered their strange hominid cousins, the robust vegetarian *Paranthropus*.

*Paranthropus* and early *Homo erectus*. Between 1.6 and 1 million years ago parts of South and East Africa were populated by two kinds of hominids. *Paranthropus* (left), a vegetarian, foraged peacefully unless disturbed, but in this imaginary encounter one is being threatened by an early true human of the species *Homo erectus*.

# Peking man and the evolution of *Homo erectus*

From 1921 up to the present day a limestone area at Zhoukoudian (Choukoutien) near Beijing (Peking) in China has produced many human fossils. These have come from cave deposits which mainly date from 250 000 to 500 000 years ago, and they represent the largest collection of *H. erectus* material known from any single site. Originally referred to as *'Sinanthropus pekinensis'*, the Chinese man of Pekin, further study of the remains showed that they were similar enough to those of Java man to belong to the same type, *'Pithecanthropus'*, ape-man, or what we now call *Homo erectus*, erect man. The majority of the remains were discovered during excavations carried out between 1928 and 1937, and they consisted of at least 14 partial skulls, 11 lower jaws, about 150 teeth and 10 fragments of the rest of the skeleton. Unfortunately almost all this material was lost in 1941, during an attempt to evacuate the fossils from China in the face of the Japanese occupation and it is still not known whether these precious bones have been destroyed. Fortunately, good documentation and replicas of the finds were made, and Chinese workers have found more material at the site since 1949.

Peking man shares a number of features with Java man which suggest they are closely related forms, but both show differences from the earlier African fossils referred by many experts to the same species. The brain of Peking man was certainly enlarged; about 25% bigger on average than the more ancient African specimens, and overlapping with the normal range in brain size

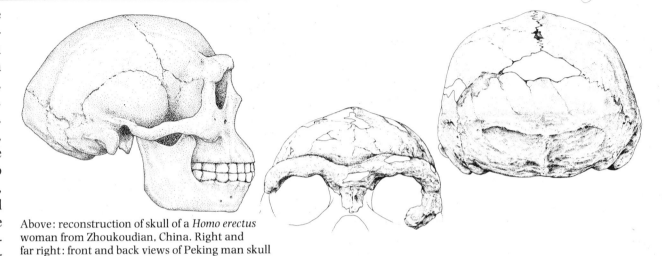

Above: reconstruction of skull of a *Homo erectus* woman from Zhoukoudian, China. Right and far right: front and back views of Peking man skull

found today. However, the brain was shaped differently and peculiarly flattened on top. The skull was strongly built at the front and back, but behind the thick browridge, Peking man had a slight development of a forehead. The teeth were relatively small but had large roots, and the lower jaw still completely lacked a bony chin. In some individuals, (perhaps females) the face and cheekbones were rather delicately built, but the overall impression is of a very muscular body.

The general assumption about the human fossils found at the main Zhoukoudian cave site is that they represent the remains of hunters and their families who camped regularly in the cave and cooked meat over open fires. Thick deposits of 'ash' in the cave are believed to represent extensive deposits of hearths, accumulated over many thousands of years, while animal bones are assumed to represent the butchered and cooked

remains of prey such as deer and rhinoceros. The human fossils may have been the remains of cannibalistic feasts since many of the skulls had their bases missing when found, which might indicate that the brains had been cut out, and some of the humans bones look burnt. However, much of this evidence is being reconsidered using new research techniques, since there are other possible explanations for the bone and ash deposits which do not involve the actions of early humans; for example, hyenas can accumulate and damage bones, while animal droppings and natural fires can create what look like ashes and hearths. Despite these doubts, stone tools and burnt bones were present in the cave, as well as a large number of human fossils. The evidence clearly indicates the presence of humans in the Zhoukoudian area over long periods of time.

Peking man. This scene is based on evidence from the Zhoukoudian site in China. It is not clear to what extent *Homo erectus* actually cooked and lived in the cave or whether the damaged human skulls found there (left foreground) indicate the practice of cannibalism.

Maurice
Wilson 1950.

# The descendants of *Homo erectus*

Africa had been eclipsed by Europe in the years up to 1920 in the quantity and importance of finds made concerning human evolution. In the following decade, however, the picture began to change with the discovery of the first australopithecine fossil from Taung (see p. 30) and of a remarkable fossil human skull from the Broken Hill mine in Northern Rhodesia (now Zambia) in 1921. The Broken Hill find, 'Rhodesian man', was made quite by chance when miners were digging out material from a cave. Over the next few years odd bones and artifacts were collected from the site, from spoil heaps, and from the homes of miners. Many other fossils must have been destroyed during the mining operations, and much knowledge of the site was lost forever.

The collection of human fossils comprises the remains of skull parts of at least two individuals and the rest of the skeleton of at least three individuals. The most important finds were of the Broken Hill skull itself, two different hip bones, a humerus (arm bone), a tibia (shin bone) and parts of femora (thigh bones). The strongly built tibia was found closest to the skull and it is long, suggesting a standing height of about 1.75 metres. One of the hip bones looks quite modern, but the other has a thickening of bone above the hip joint like that found in much older fossils from Africa and Europe.

The Broken Hill skull is one of the best preserved of all fossil human relics and gives us a good idea of what the powerful face of its owner, probably a young man, looked like. The eyes must have been overshadowed by a very prominent browridge,

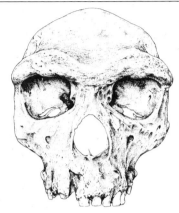

Facial view of the Broken Hill skull from Zambia

one of the largest known in any fossil, while the nose was broad, not particularly long, and quite flat. The upper jaw was large, but the teeth were not big, and not healthy. In fact the Broken Hill man suffered from the first serious case of tooth decay known in an early human. He may also have had a bone tumour in the side of his head, and these health problems could have been the cause of his death. His brain, 1280 ml in volume, was only a little smaller than the modern average and his skull bones show combinations of characters found in both *H. erectus* and modern skulls.

Elsewhere in Africa, other fossil skulls and jaws have been found which probably represent the same people as the Broken Hill skull. Despite the differences these all show from modern remains, they are usually regarded as representing a very early type of our own species *Homo sapiens*, wise man. A skull even more massive than that of Rhodesian man, found at Bodo in Ethiopia and dated at about 400 000 years old, may be an early

representative of the same people. Another skull from Elandsfontein (Saldanha, South Africa) at the other end of the African continent probably represents a female of the same type, and was found associated with handaxes perhaps 350 000 years old. Her skull is smaller than the Broken Hill and Bodo ones, with a less prominent browridge.

In Europe, a similar group of fossils probably represent the European relatives of these descendants of *H. erectus*. A skull, jaws, teeth and pelvis from Arago in France, a skull from Petralona, Greece, a lower jaw from Mauer, West Germany and skull fragments from Bilzingsleben, East Germany and Vértesszöllös in Hungary give us a glimpse of the early Europeans of 300 000 to 500 000 years ago. Depending on their location and the materials available, these people lived in cave entrances or in simple huts which they built from wood and leaves. As well as handaxes and flake tools (made from flakes of rock, for example flint), a few implements of bone and wood, including a spear, have also been preserved from this time.

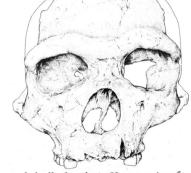

Reconstructed skull of archaic *Homo sapiens* from Arago

'Rhodesian man'. An African family of about 200 000 years ago. The adult man is reconstructed ▷ from the Broken Hill remains (Zambia) while the woman's head is partly based on the Saldanha skull (South Africa). No fossil remains of children of this type have yet been found.

# Swanscombe, and human evolution in Europe

Throughout this century the gravel pits around the village of Swanscombe in Kent have produced prodigious quantities of stone tools, especially handaxes, of the Acheulian industry (first named at St. Acheul in France). It was in the 'Middle Gravels' that the first part of a human skull which might represent one of the makers of these handaxes was discovered in 1935. This find was of an occipital bone (the back and base of the skull) and was followed a year later by the discovery of the left parietal bone (side wall) which fitted together with it perfectly. Finally, in 1955, the right parietal bone of the same skull was found, making up the complete back half of the skull of an adult.

The fossil skull did not show the markings of strong muscles, suggesting that it may have belonged to a female, although the bones are very thick. The brain size, about 1325 ml, was indistinguishable from our own, and overall the specimen appears quite modern when compared with a *Homo erectus* skull. However, this is a little deceptive, as there is no evidence of the front part of the Swanscombe skull, which may well have had large eyebrow ridges and a big face. Despite the modern appearance, there are some unusual or primitive features on the bones preserved at Swanscombe such as the flattened top of the parietals, the broad base, down-turned edges and central pit of the occipital bone. At first, the modern-looking shape and gracility of the skull led many experts to classify it as a modern *H. sapiens* or as related to the fraudulent Piltdown skull from Sussex, which we now know consisted of an artificial combination of a recent human skull and the jaw of an orang-utan. Recent research instead indicates closer links to the Neanderthals of Europe (see p. 44), who share the unusual Swanscombe skull features mentioned earlier, some of which seem to relate to distinctive muscles of the neck.

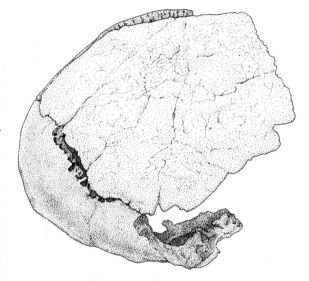

Side view of the preserved rear of skull from Swanscombe, usually assigned to archaic *Homo sapiens*

The idea that the Neanderthals might trace their ancestry back at least to the time of the Swanscombe fossil (estimated at about 250 000 years old) is supported by other evidence such as more recent finds from Biache in France and Pontnewydd in Wales. At Biache, a site somewhat younger than Swanscombe, a very similar partial skull with fragments of the upper jaw was found in 1976. The bones of the skull are much thinner than those from Swanscombe and they are noticeably more Neanderthal-like in their rather spherical shape when viewed from behind, and the exaggerated projection of the occipital region. The Pontnewydd remains, which represent at least three different people (a child, an adolescent and an adult) only consist of teeth, jaw fragments and a single bone from the spinal column. They were found in a cave in North Wales during excavations begun in 1978. Although some specimens are too incomplete to be distinguished from those of modern people, some of the teeth are remarkably like those of Neanderthals in their development of taurodontism (unseparated tooth roots with large pulp cavities). The Pontnewydd remains are of about the same age as the Swanscombe bones, more than 200 000 years old, and may represent a similar early British population of proto-Neanderthals living at the margins of the inhabited world.

The picture we can build up of life in Europe at this time indicates that these people were opportunistic in the way that they exploited the environment. They were probably very mobile, camping briefly near water and food supplies, before moving on again. They could have taken medium sized game, such as deer, by hunting, and larger game where they could scavenge carcasses produced by natural deaths, accidents or carnivore kills. However, collecting wild fruits, plants and vegetables would still have been a very important source of food for the Swanscombe people.

Swanscombe. A deer hunt on the banks of the river at Swanscombe, 250 000 years ago. A wooden spear has brought down a specimen of the extinct Clacton fallow deer, *Dama dama clactoniana*. The skull of a wild ox lies in the foreground. ▷

# The Neanderthals

The people who immediately preceded the first modern-looking inhabitants of Europe were the Neanderthals. The typical or 'classic' Neanderthals lived from about 70 000 to 35 000 years ago during the first part of the last ice advance in Europe although, as we have already seen, their ancestors can be recognized long before this last phase. At the peak of their evolutionary success the Neanderthals also lived across Asia, in the Middle East, and as far as Uzbekistan in the Soviet Union.

Like earlier humans, the Neanderthals had skeletons which emphasized physical strength, with thick bones, big joints and heavy musculature, even in females and children. They probably grew up and matured faster than we do, but were short and stocky in build, weighing up to 70 kg but measuring only about 1.65 metres in height. Their physique was probably adapted to the cold conditions in which many of them lived, and they had a large trunk and short legs, especially the shin bones. Their faces may also have been shaped through climatic adaptation, since their noses were unusually large and projecting and their cheek bones were swept back at the sides. Neanderthals also differed from earlier humans in their overall skull shape and their enlarged brain, which averaged about 1450 ml, even larger than the modern average. We do not know whether the large brain of the Neanderthals reflected a high intelligence, the need to control a large and highly muscled body, or both.

Most Neanderthal remains have been found in caves because that is where the Neanderthals buried their dead; they are the first people known to have done this. We are not sure whether the Neanderthals had primitive religious beliefs, but some burials seem to have been accompanied by some sort of ritual. In one cave in Italy a Neanderthal man's skull was found in a ring of stones; in another cave in Iraq it is claimed that a man was buried with flowers over him. Thus some aspects of Neanderthal behaviour appear very like that of modern humans, but in other ways they may have been markedly less efficient than modern hunter-gatherers in the way they exploited the environment, with a reliance on opportunistic foraging and brute strength to achieve limited aims, rather than on more careful long term planning and a good understanding of their surroundings.

From the time of the first recognized Neanderthal find from Germany in 1856 (even earlier finds had been made in Belgium and Gibraltar) there has been much dispute about whether the Neanderthals were ancestors of modern people in Europe or elsewhere. Much of the argument has hinged on proving or disproving the existence of other possible ancestral types from elsewhere, and from recognizing 'intermediate' characters in late Neanderthal or early modern fossils. In Europe we now know that some Neanderthals were still living there unchanged when the first modern people appeared, so they cannot have been our ancestors, at least not in that area or in that exact form.

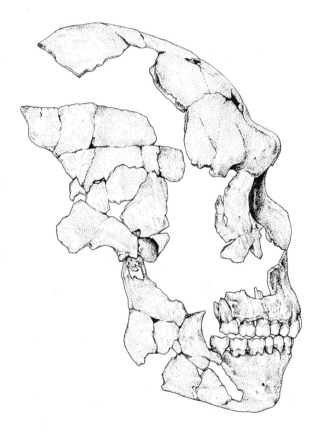

A late Neanderthal skull and lower jaw from St. Césaire in France. About 35 000 years old

Neanderthal man. Reconstructed from remains from various Gibraltar caves, this group is outside Gorham's Cave, looking north and east to the shrunken Mediterranean, lowered by loss of water locked up in ice caps 50 000 years ago. One of the hunted birds is a Great Auk.

# The Cro-Magnons

About 35 000 years ago a new type of early human appeared in Europe named 'Cro-Magnon' after a French cave which produced some of the earliest remains in the last century. These peoples were much more like the present inhabitants of Europe than the earlier Neanderthals were; in fact their main differences were a slightly heavier build and a slightly larger brain than the modern average.

The Cro-Magnons were tall and straight-limbed. The men averaged over 1.8 metres and the women over 1.65 metres in height, and both sexes would have been well muscled by modern standards. Their physique was rather like that of people who live in warm climates today, despite the fact that the Cro-Magnons lived in conditions every bit as cold as those endured by the Neanderthals. This is perhaps an indication that the Cro-Magnons originated in warmer climates, as well as being well adapted culturally to cope with the unfavourable climate.

These people made a more intensive use of their environment than the Neanderthals did, and were probably better at planning their activities, including the hunting of large game such as mammoth, woolly rhinoceros, giant oxen, horse and reindeer. Their hunting weapons included hafted spears, which would have been aided in some cases by the spearthrower to give a better range and force of throw, and probably traps. Some Cro-Magnons were also proficient at fishing, using lines, hooks, harpoons and probably also nets and boats. They produced stone tools made from narrow blades, which could then be further

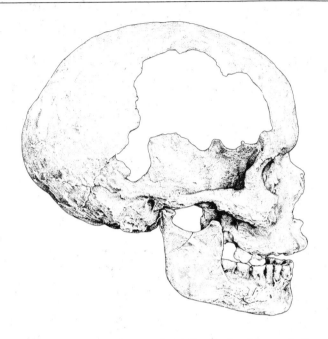

Two Cro-Magnon specimens from Czechoslovakia – a skull from Mladeč and a lower jaw from Brno. The Mladeč cranium is probably more than 30 000 years old.

modified to create specialized artefacts such as scrapers, engraving tools, knives and borers. The various industries made by the Cro-Magnons at different times and places, such as the earlier Aurignacian 30 000 years ago or the later Magdalenian of 15 000 years ago, are grouped under the collective name Upper Palaeolithic (Upper Old Stone Age). This cultural stage was also characterized by the proliferation of cave art.

The origin of the Cro-Magnons is still a subject of dispute. Some experts believe that they could have evolved fairly rapidly from the preceding Neanderthals, given strong enough selection pressure for 'modern' characteristics and for the behavioural and cultural changes which led up to the Upper Palaeolithic. Support for this idea comes from suggested evolutionary changes in the direction of the Cro-Magnons found in some Neanderthals, supposed Neanderthal-like characters in some Cro-Magnon fossils, and possible cultural links between the two peoples. Other scientists believe that the differences between the two groups were too profound to allow an evolutionary transition between them in the short time available, and argue that the Cro-Magnons were the descendants of immigrants from outside the European continent. They point to the existence of earlier modern-looking people in South West Asia and Africa, who lived at the same time as the Neanderthals of Europe. These peoples had probably evolved from the earlier non-Neanderthals of Africa, and spread out of that continent more than 50 000 years ago. They reached eastern Europe by 40 000 years ago where they began to displace the occupying Neanderthals, and similar groups spread to the Far East from where they became the first colonizers of Australia and the Americas. From this reading of the evidence, the Neanderthals could not have been our ancestors, and the Cro-Magnons were just one of many similar early modern groups who spread around the world during the last 50 000 years. The variation we see in the peoples of the world has mainly originated during this period.

Cro-Magnon man. The old man on the right is reconstructed from the original French Cro-Magnon fossil, others from the more rugged East European skeletons. Cave paintings were usually much further from the entrance than shown here.

Wilson 1951

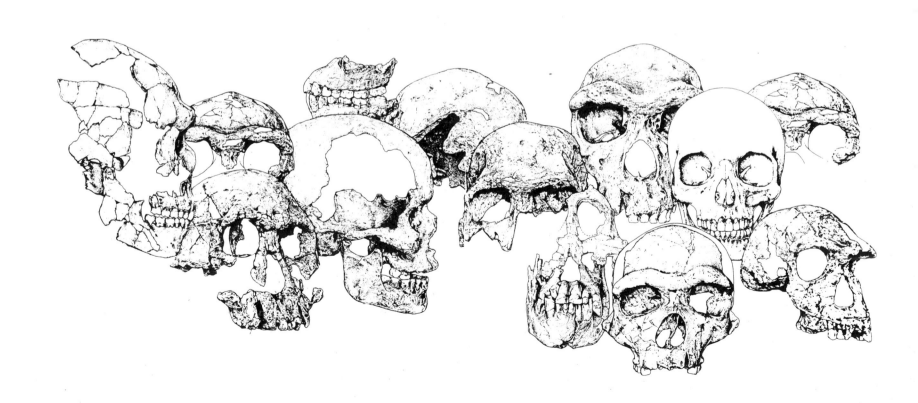